MW01248493
27A

Rodney Rootle's GROWN-UP GRAPPLER and Other Treasures from The Museum of OUTLAWED INVENTIONS

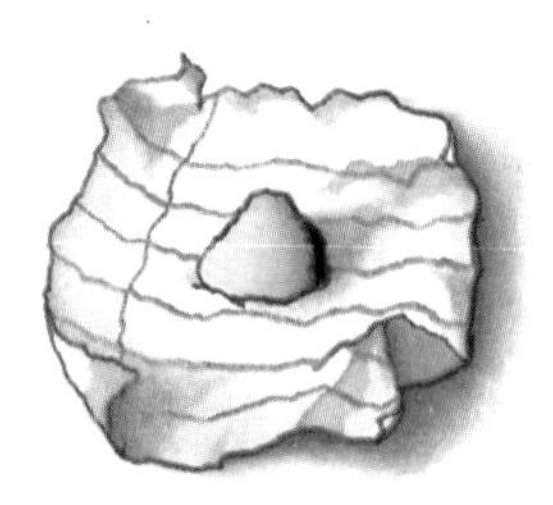

ATLANTIC–LITTLE, BROWN BOOKS
ARE PUBLISHED BY
LITTLE, BROWN AND COMPANY
IN ASSOCIATION WITH
THE ATLANTIC MONTHLY PRESS

BASED ON AN IDEA BY JEREMY BEADLE.
INVENTION IDEAS BY CHRIS WINN,
WITH SOME IDEAS BY JEREMY BEADLE.

Library of Congress Catalog Card No. 82-082078
First American Edition

ISBN 0-316-94752-0

Designed by Nick Thirkell
Printed and bound in Belgium by Henri Proost
For the Publishers Pepper Press, 1982
An imprint of Evans Brothers Limited, London

CHRIS WINN
JEREMY BEADLE

An Atlantic Monthly Press Book
LITTLE, BROWN AND COMPANY
BOSTON

INTRODUCTION

Deep below the busy streets of Central London lies a top secret bunker. It is protected by an elaborate security system so efficient no one has ever penetrated its hidden secrets. This subterranean maze houses the most extraordinary museum in the Western World: the legendary Museum of Outlawed Inventions.

Here is a complete collection of forbidden devices, each and every one designed by kids. It doesn't take long to see why the inventions have been banned, for every one is a lazy child's dream-machine. Only recently discovered, this staggering display of sensational ingenuity contains devices guaranteed to make any child's life bliss.

After careful consideration and against our better judgement, we have decided to reveal for the very first time what adults over the years have been determined to keep hidden. We have taken great risks to bring the achievements of these young inventors to you.

Devious and daring as these inventions are, they are only a beginning. The legendary Museum of Outlawed Inventions is ever eager to expand its collection. Each new generation brings its own extraordinary imagination and skill to the task of adding to this incredible assortment of mind-boggling gadgets.

Surely you have an outrageous invention of your own to contribute?

3.
2.
4.
1.
Guard
5.
Wuf!
Guard Dog

GHOUL ROOM
Hall of Wib-Webs
Goblinarium
Mechanical Skellingbones
Lunar Powered Snilderwizzles
Emergency Resuscitation Unit
Toilets
SPOOKY INVENTIONS
A selection of inventions designed to scare parents, relatives and babysitters is shown here, with a view looking across the main foyer of the museum to the Ghoul Room. Also in the picture is one of the museum's new, specially trained night guards, who is doing her first night-shift.
1. Sofa-Coffin
This fine upholstered sofa was made by Marmaduke, aged 10, as a present for his grandfather, the late Baron von Snivel-Pilchard. A harmless old man, of short memory and aching limbs, he sat down on the sofa, and lured by its luxurious comfort, leaned back. Immediately, activated by a hidden mechanism, the cushions sank, the sides went up, and the top came down. Thus imprisoned, the late Baron was deafened by a funeral dirge from built-in loudspeakers, and his feet were tickled by revolving parrots' feathers. Much later, having managed to release himself from the coffin, the late Baron was so shaken in memory and limb, that seeing a luxurious sofa beside him, he at once sat down on it again.
2. Aspidistropus Strangulosus
An accidental cross between a houseplant and an octopus.
3. Ghoul Mirror
One of a pair of 18th century roll-top ghoul mirrors.
4. Headless Ghost Standard Lamp
Hand-painted silk drape unfurls from rim of lampshade thirty minutes after switching on the lamp.
5. Pet Dog Werewolf Mask
With slobberized gnasheration.
6. Skeleton Rug
A footstep-activated, portable pop-up.
7. Clockwork Boneshaker
Climbs up walls and rattles on windows.
6.
7.

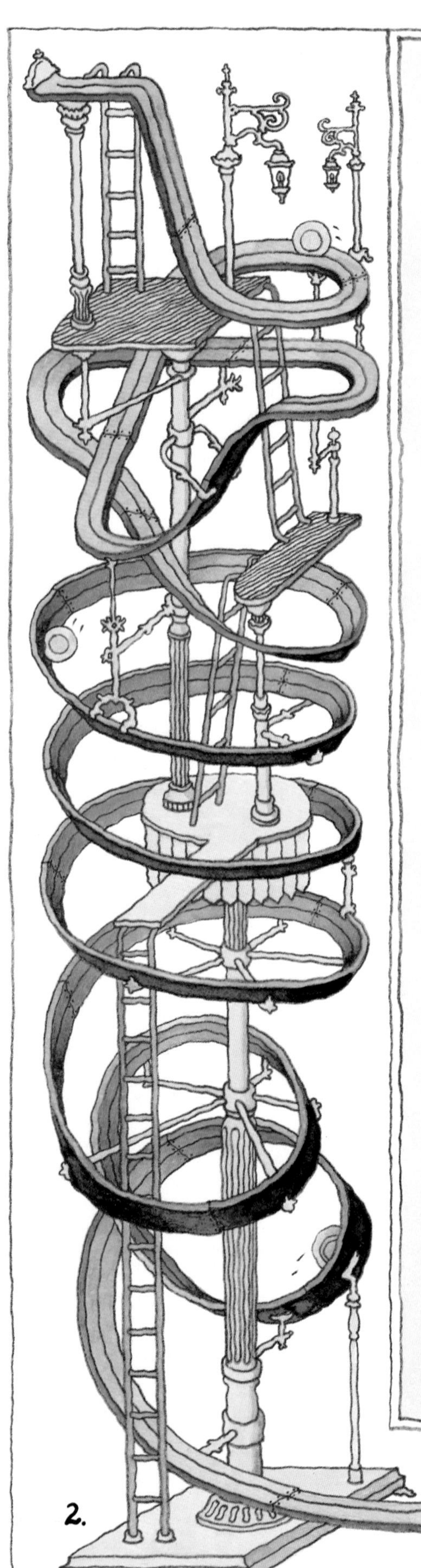

ANTI-ADULT INVENTIONS

1. Grown-Up Grappler

This strange apparatus was designed and built in 1967 by Rodney Rootle, then aged 10, as a way of protecting himself from the soppy greetings and embraces of his innumerable aunts and great-aunts. It was apparently made from a very expensive 18th century table belonging to his mother, an old dustbin, a plastic food cover, some vacuum cleaner hose, and a pair of rubber gloves. The Grown-Up Grappler is on temporary loan to the museum, and in his description of the apparatus Mr Rootle mentions the following features: an automatic kisser, with 'realistic' saliva spray; fully extendible hugging arms; a silly questions answering unit, with pre-recorded tapes for each aunt; a water cannon, a bubblegum flicker, and an off-putting smelly sock. (These last three were presumably intended to counter the advances of dangerously affectionate aunts.) Mr Rootle adds that he was a boy of restless imagination but little practical ability. His apparatus is perhaps the least successful invention in the museum.

2. Victorian Plate Smasher

This elaborate plate smasher, one of several similar devices in the museum's collection, was the brainchild of Lady Agatha FitzThrowen, youngest daughter of the Duke of Tantrum. Every summer the Duke and his *entourage* abandoned Tantrum Castle for Europe, leaving Lady Agatha at home with her nanny. As a protest against this annual dirty trick, Lady Agatha would have the Plate Smasher set up on the castle lawns, and by the time the Duke returned from his tour, the family china was elegantly and systematically dispatched.

3. Ye Olde Relative Flattening Bedde

Designed in the reign of Elizabeth I by Francis Grope, a particularly nasty 5 year old boy. Visiting relatives were threatened nightly by 4½ tons of concealed granite, controlled with pulleys and winches by Francis in his bedroom next door.

4. Super-Sticky Snot Formula

A page from the coded notebook of Alfie Greenstough, showing his famous recipe for extra-sticky snot, which was guaranteed to stick to the moustaches of sleeping dads and uncles. Specimen of finished product also shown.

5. Automatic Clothes-Scattering Chair

Powerful centrifugal motion scatters neatly folded clothes. For placing in goody-goody cousins' rooms who come to visit.

1.
3.
4.
5.

FOUL FOOD INVENTIONS

Down in the basement, deep in the bowels of the museum, is the Department of Inventions for Dealing with Foul Food. Some specimens from this department are shown here.

1. De Luxe Yuck Gulper

The Yuck Gulper was invented in 1952 by Graham Runge, then age 9, as a desperate answer to the problem of eating up his vegetables. Standard features were protective gloves, plate-glass safety canopy, and steel-tipped pulverising *Dyna*Chompers.
The de luxe model shown here also has a self-fuelling gas converter (1a), which turns the pulverised vegetable matter into deadly poisonous and highly explosive Foul Food Gas [$PoOH_3$].

2. Brussel Sprout Time Transporter

This machine is powered by $PoOH_3$ from Invention 1, and is designed to send offending food particles through time and space. The Time Transporter is, in fact, a modification of the Diesel-Powered Centrifuge of Miss Sylvia Smalls, Headmistress of St. Werburgh's Preparatory School for Boys, who designed it in 1887 for rapid drying of soggy sports kits.

3. Inter-Planetary Sprout Monitoring System

Sprouts from Invention 2 can be guided to the destination of your choice with this very recent product of sprout disposal technology.
In the picture three typical sprouts can be seen at various stages of their journey to Mars.

4. Pocket Gristle Detector

This was invented by Philip Stakes, aged 13, during the lunch-hour of 14th May 1979, after a particularly bad school stew.

Also in the picture are some Remote-Controlled Plastic Maggots (5), and a passage leading to the Stink-Bomb Bottling Plant (6).

IMPORTANT NOTICE

Visits to the department are by special permission of the Head Keeper only. Visitors are reminded that the museum can accept no responsibility for any sickness, fainting or undesirable collywobular activity caused by excess inhalation of the noxious vapours in the department. Protective clothing is provided and must be worn at all times. Allow at least two hours on exit for processing through the air locks and de-contamination chambers.

2.
3.
6.
4.
6.03
SPROUT·1·
ARRIVED
SPROUT·2·
IN ORBIT
SPROUT·3·
READY
SPROUT 3
READY
R.P.M
uction
control
FF
LOW
AST
OSE PEG
ispenser
Small
Medium
Large
TEMPERATURE
PRESSURE
NORMAL
OFFENSIVE
STINKERATION

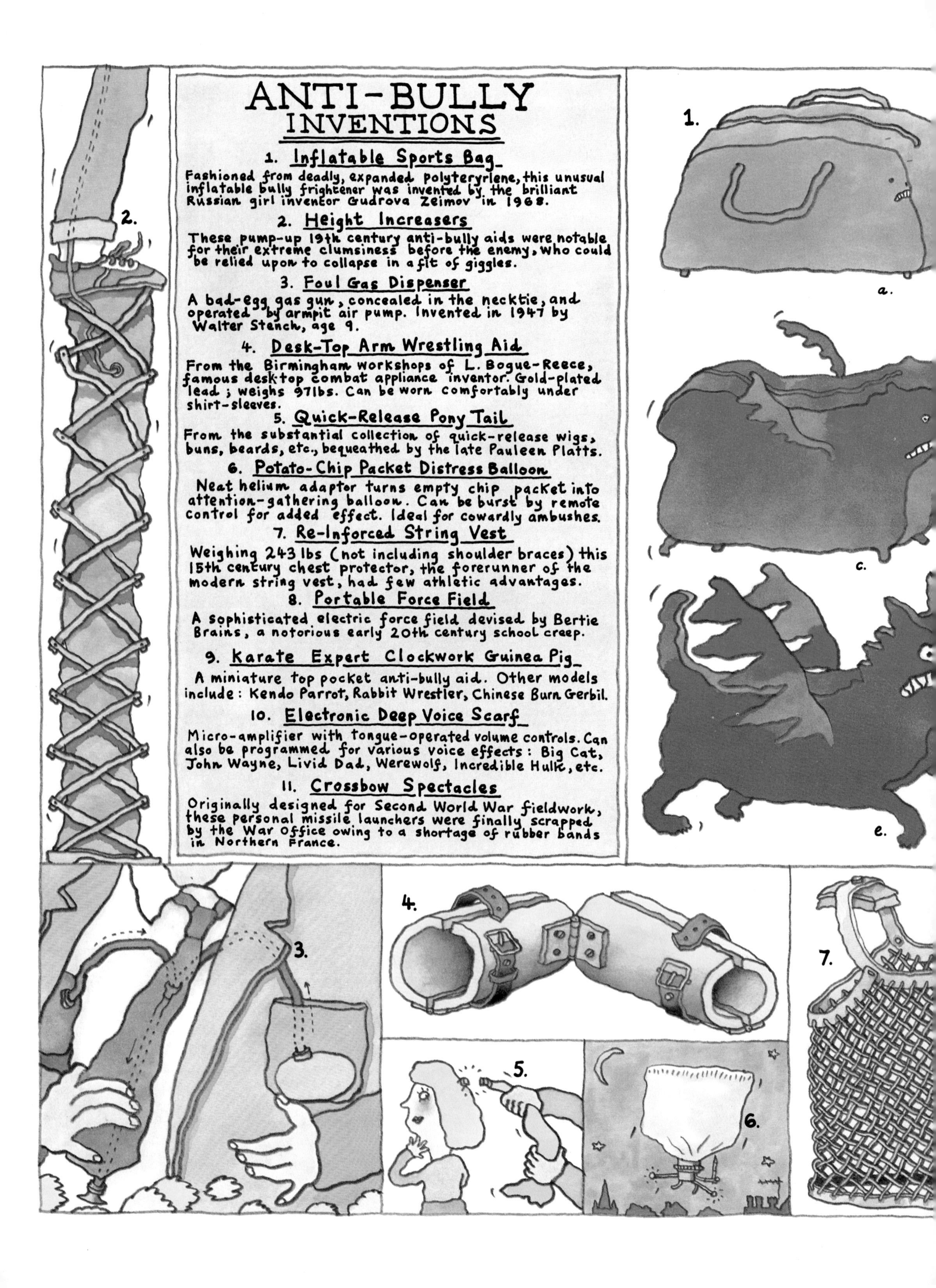
ANTI-BULLY INVENTIONS
1. Inflatable Sports Bag
Fashioned from deadly, expanded polyteryrlene, this unusual inflatable bully frightener was invented by the brilliant Russian girl inventor Gudrova Zeimov in 1968.
2. Height Increasers
These pump-up 19th century anti-bully aids were notable for their extreme clumsiness before the enemy, who could be relied upon to collapse in a fit of giggles.
3. Foul Gas Dispenser
A bad-egg gas gun, concealed in the necktie, and operated by armpit air pump. Invented in 1947 by Walter Stench, age 9.
4. Desk-Top Arm Wrestling Aid
From the Birmingham workshops of L. Bogue-Reece, famous desk-top combat appliance inventor. Gold-plated lead ; weighs 97lbs. Can be worn comfortably under shirt-sleeves.
5. Quick-Release Pony Tail
From the substantial collection of quick-release wigs, buns, beards, etc., bequeathed by the late Pauleen Platts.
6. Potato-Chip Packet Distress Balloon
Neat helium adaptor turns empty chip packet into attention-gathering balloon. Can be burst by remote control for added effect. Ideal for cowardly ambushes.
7. Re-Inforced String Vest
Weighing 243 lbs (not including shoulder braces) this 15th century chest protector, the forerunner of the modern string vest, had few athletic advantages.
8. Portable Force Field
A sophisticated electric force field devised by Bertie Brains, a notorious early 20th century school creep.
9. Karate Expert Clockwork Guinea Pig
A miniature top pocket anti-bully aid. Other models include: Kendo Parrot, Rabbit Wrestler, Chinese Burn Gerbil.
10. Electronic Deep Voice Scarf
Micro-amplifier with tongue-operated volume controls. Can also be programmed for various voice effects: Big Cat, John Wayne, Livid Dad, Werewolf, Incredible Hulk, etc.
11. Crossbow Spectacles
Originally designed for Second World War fieldwork, these personal missile launchers were finally scrapped by the War Office owing to a shortage of rubber bands in Northern France.
1.
a.
c.
e.
2.
3.
4.
5.
6.
7.

b.
d.
f.
11.
10.
SPEAK HERE
ON-OFF
8.
9.

3.
4.
5.
GUARD
1.
6.
APPROVED

SUBMARINE INVENTIONS
SUBMARINE
INVENTIONS
1. The Underwater Fiddlegrippit
When it was invented in 1892, people thought the Fiddlegrippit, with its unique hydraulic scuttle motion, was for serious undersea exploration. It was not until 1979 that it was found uprooting anchors, twisting ships' propellers and wrestling with submarines. Still at the controls, its inventor, Oscar Hooligan, now aged 96, was arrested and the Fiddlegrippit handed over to the museum.
2. Bumflough's Electric Shark
A remote-controlled treasure-seeker invented by Billy Bumflough, the notorious boy pirate.
3. Pedal Squid
This has telescopic suction tentacles for pulling down bathing suits. It was invented in 1953 by Doris Dimple, aged 9.
4. Gluey Jellyfish
Also designed by Doris Dimple, these stuck fast to people's bottoms, and could be fired from the Pedal Squid.
5. Homing Barnacles
These were radio-controlled and gave off a special perfume irresistible to ordinary barnacles. This enabled their sinister inventor, S. Q. Welch, to control barnacles by the billion, with disastrous results to passing ships and swimmers.
6. Exploding Sea Cucumbers
Designed to clear swimming-pools instantaneously. Invented by Elmer Gork, aged 14, a very lazy pool attendant and ex-pirate.
2.
FAST

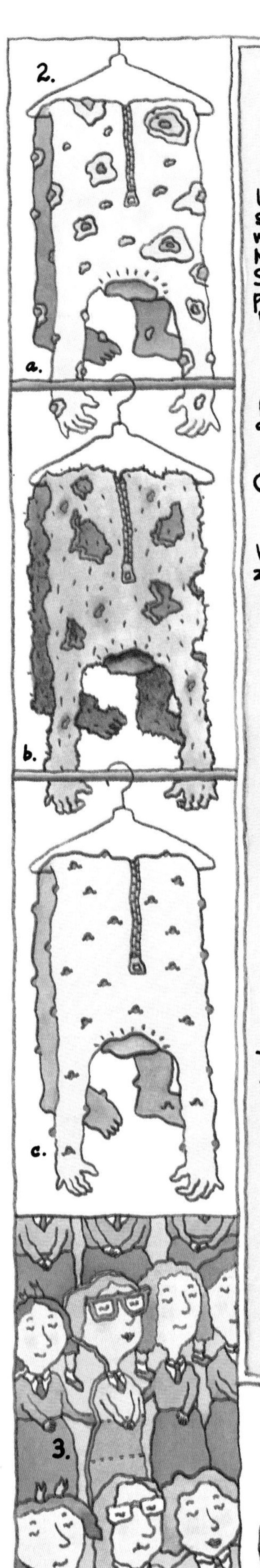

SCHOOL INVENTIONS

1. Headmaster Tracking System

Unknown to most people the very first satellite in space was the Inter-Classroom Headmaster Tracker, which was launched secretly from the roof of the National Association for Schoolboy Aids in 1952. Signals from the Tracker (a) were relayed to personal video receivers (b), which gave instant warning of approaching danger.

2. Lurgy Suits

Exotic skin disease outfits designed to repel parents and teachers. Made from heavy-duty Pustulon.
a. Brazilian Green Boil b. Hairy Mildew c. Itchy Pimple

3. Assembly Stand-In

Cardboard look-alike with optional roll-call answerer.

4. Silent Potato-Chip Eating Mask

With powerful noise-baffles each side of chewing zone. Silent to a level of 5.6 sonic munches.

5. Fake Pencil-Case Drink Dispenser

6. Excuse Letter Writer

This portable electronic Excuse Letter Writer was invented in 1977 by Alex Plain, aged 12. The creative brain of the machine is a powerful 2000E Micro *Diddler*, which can cope with anything from a simple dentist note to a 16 page permanent sports excuse. Other features include a fib counter, cringe-level metering, and a parental handwriting specimen entry.

7. Automatic Chalk Returner

A desk-top anti-teacher aid, housed in a realistic lunch box, and featuring electronic chalk detection, split-second countdown and high-velocity 'ping-pong' action return.

8. The Art of Hypnotising Teachers

These simple-minded creatures respond readily to hypnotic suggestion, as explained in this informative volume by the schoolboy master of the art, Alistair Attew.

9. Classroom Handypack

A miniature satchel containing instant faint nose drops, Miss Rique 'Boys off' perfume, and anti-chalk blackboard spray.

10. Video Desk

From the museum's large collection of customized desks, this model features a concealed television with under-desk bubblegum blob controls.

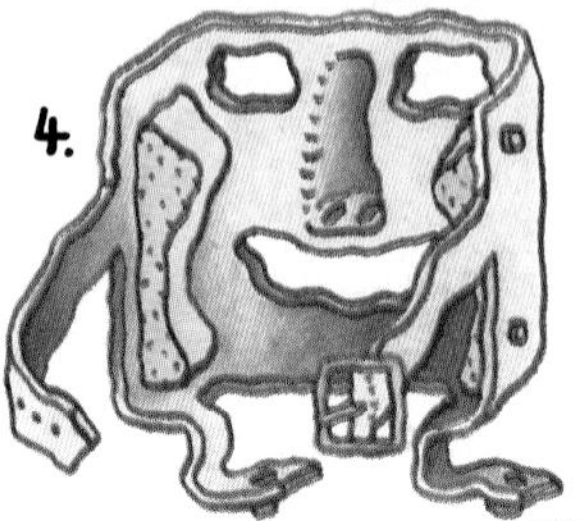

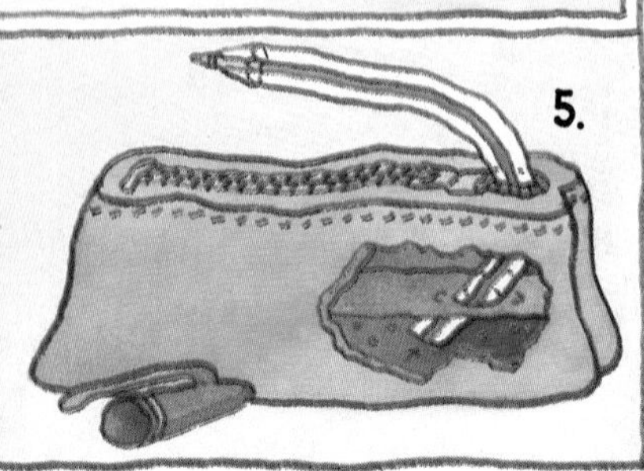

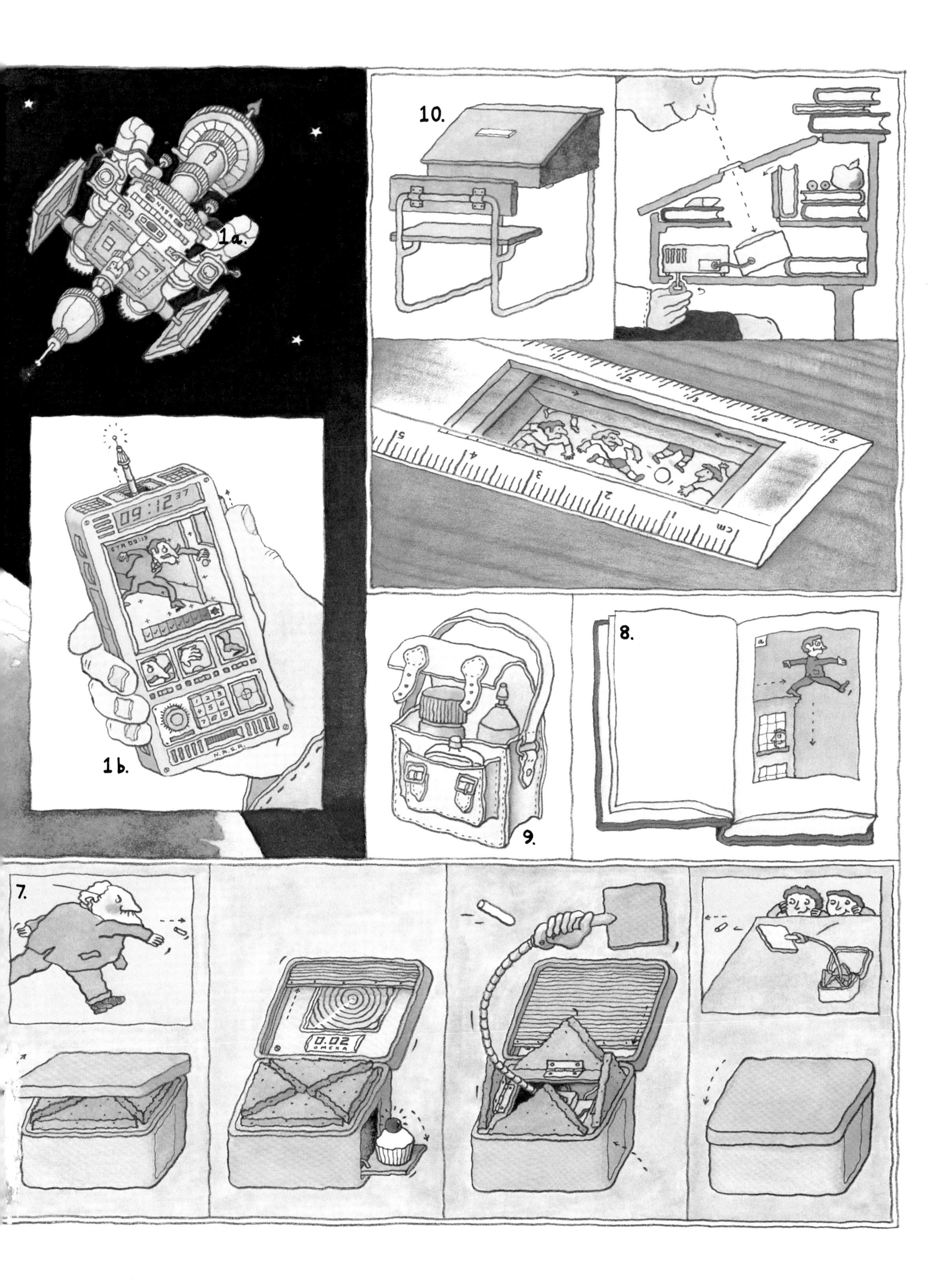

1a.
1b.
09:12
N.A.S.A.
10.
cm
8.
9.
7.
0.02

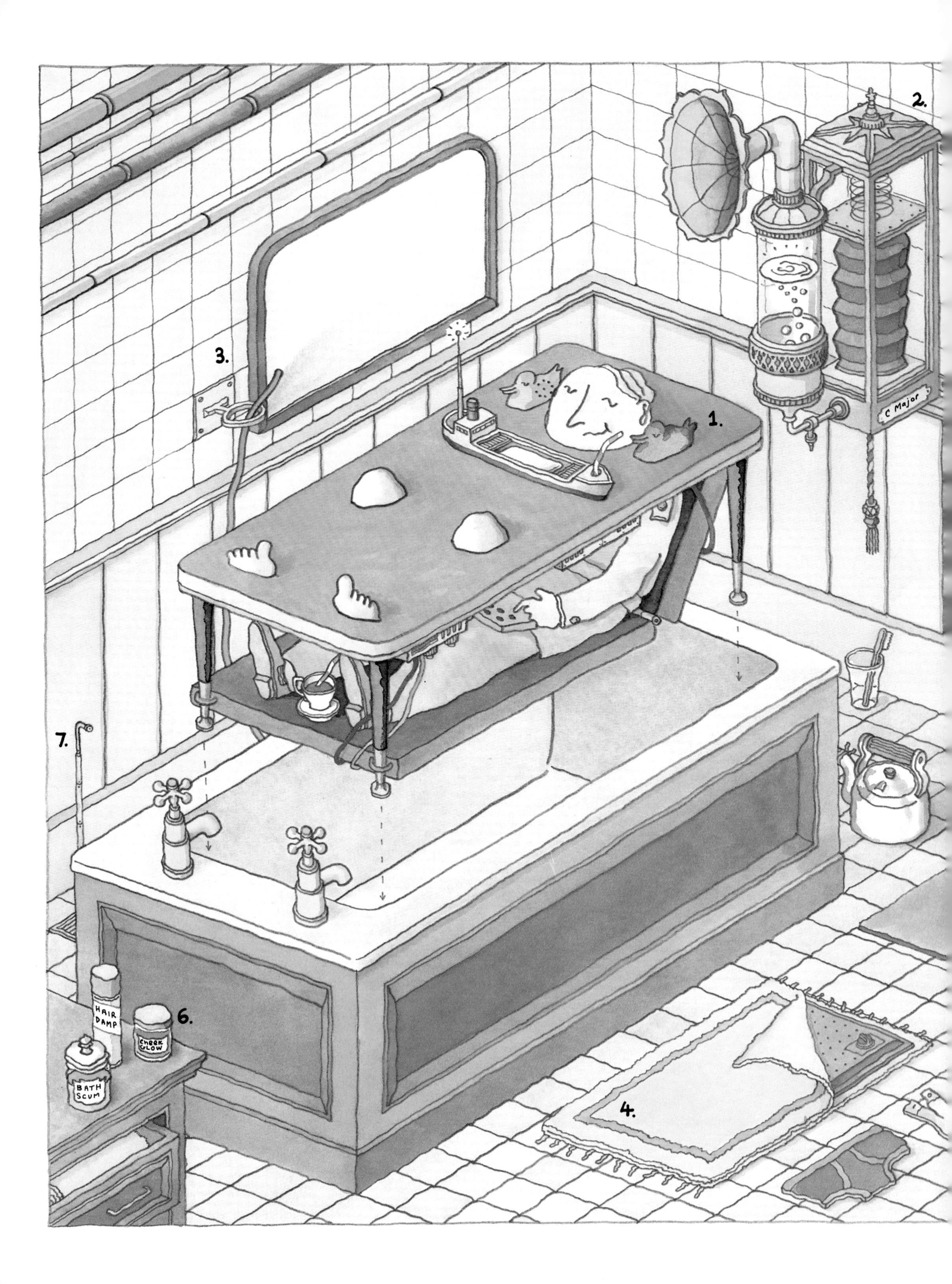

1.
2.
3.
4.
6.
7.
C Major
HAIR DAMP
CHEEK GLOW
BATH SCUM

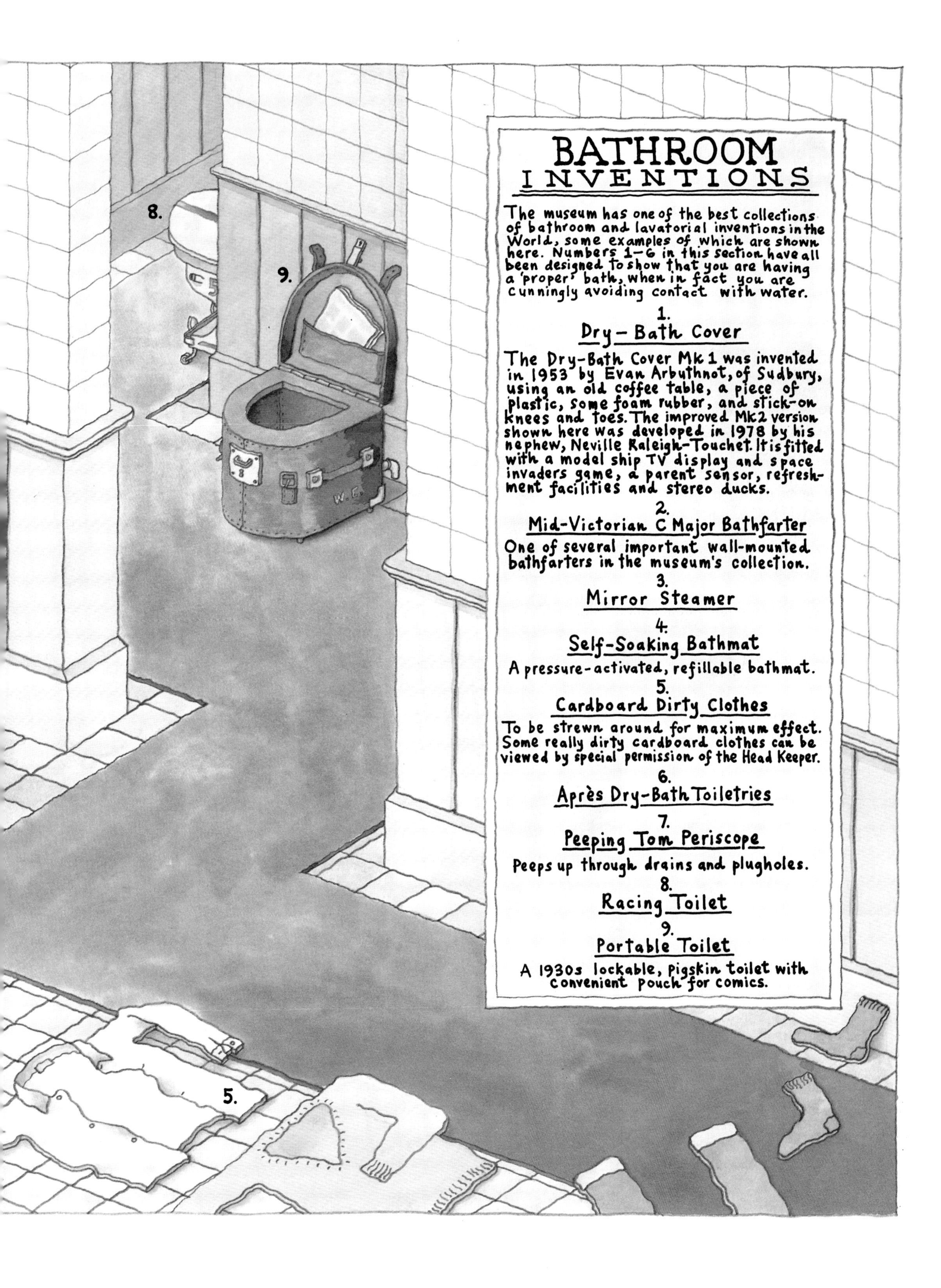
BATHROOM
INVENTIONS
The museum has one of the best collections of bathroom and lavatorial inventions in the World, some examples of which are shown here. Numbers 1–6 in this section have all been designed to show that you are having a 'proper' bath, when in fact you are cunningly avoiding contact with water.
1.
Dry – Bath Cover
The Dry-Bath Cover Mk 1 was invented in 1953 by Evan Arbuthnot, of Sudbury, using an old coffee table, a piece of plastic, some foam rubber, and stick-on knees and toes. The improved Mk2 version shown here was developed in 1978 by his nephew, Neville Raleigh-Touchet. It is fitted with a model ship TV display and space invaders game, a parent sensor, refreshment facilities and stereo ducks.
2.
Mid-Victorian C Major Bathfarter
One of several important wall-mounted bathfarters in the museum's collection.
3.
Mirror Steamer
4.
Self-Soaking Bathmat
A pressure-activated, refillable bathmat.
5.
Cardboard Dirty Clothes
To be strewn around for maximum effect. Some really dirty cardboard clothes can be viewed by special permission of the Head Keeper.
6.
Après Dry-Bath Toiletries
7.
Peeping Tom Periscope
Peeps up through drains and plugholes.
8.
Racing Toilet
9.
Portable Toilet
A 1930s lockable, pigskin toilet with convenient pouch for comics.
8.
9.
W.C.
5.

KEVIN JONES'S

HOMEWORK INVENTIONS

Kevin Jones was clever and could do homework in ten minutes, but his mother believed in three solid hours a night. At the age of 6, bored with handwriting practice and confined to his bedroom, he invented the Squeaky Pen Machine (1). This convinced his mother that he was hard at work when she listened at his door. Meanwhile, he could get on with the serious business of playing with his white mice. A year later, just as Mrs Jones was becoming suspicious, he built the Occasional Pen Tapper (2).

By the time he was 10, Kevin was so fed up with hours of French and Physics, he started tinkering with his dad's stereo, and made the Loud Page Turner (3), and the Head Scratching Device (4). Then his mother was so impressed by his intense mental activity, she stopped listening at the door altogether.

But about this time his sister Priscilla got suspicious, and he was obliged to invent the Sneak Sensor (5). This had an ultra-sensitive Sniffer (6), which detected Priscilla's creepy feet as she approached.

One Monday evening, instead of doing History, Kevin made the Bedtime Warner (7). This was a straight conversion of his dad's digital alarm clock.

By the time he was 12, Kevin was so bored with his homework, he decided to invent the Special Text Converter Mk I (8), which turned dull school books into exciting films.

Finally, when he was 13, driven to distraction by endless dreary essays, Kevin knew the time had come to solve the homework problem once and for all. And so, during the summer holidays, he designed and built the Homework Processor 6000E Superette (9), which took full charge of homework, and always got Grade 'A's.

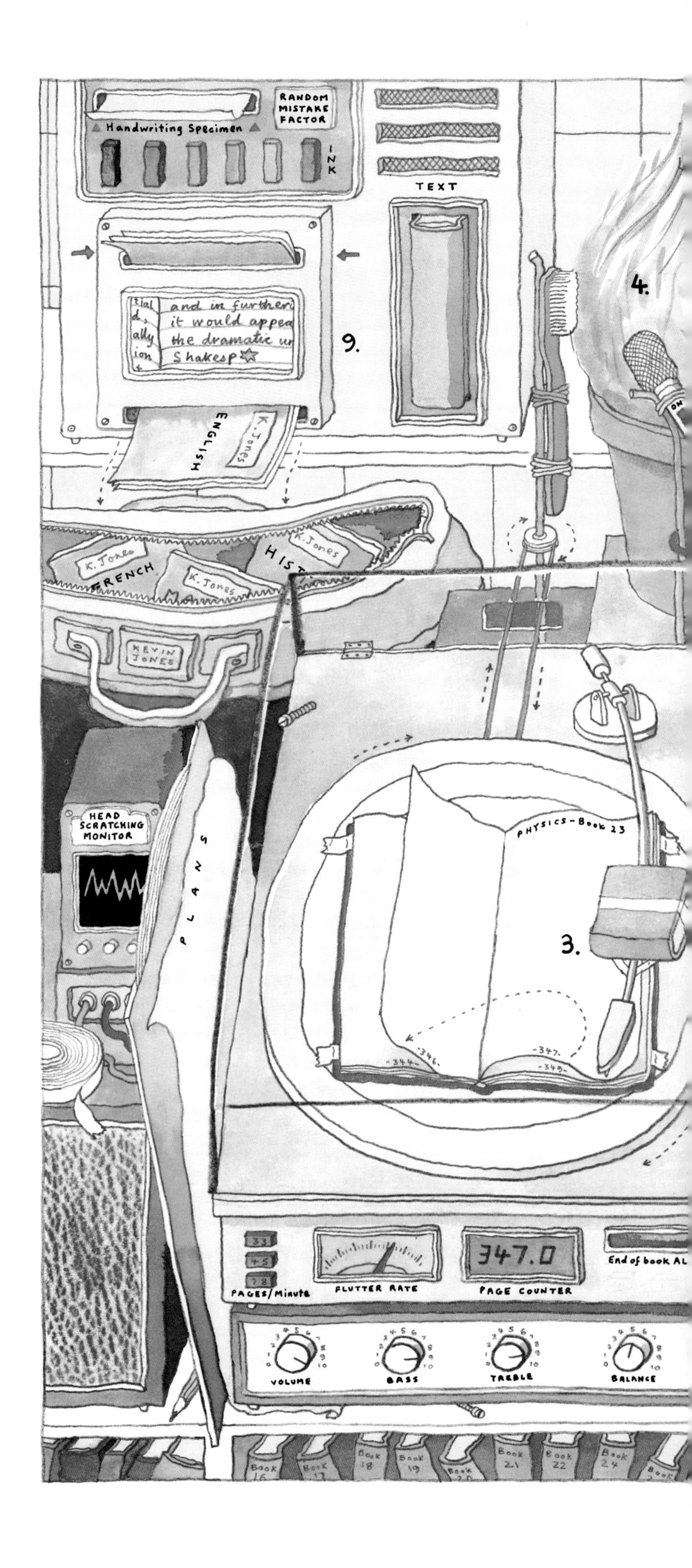

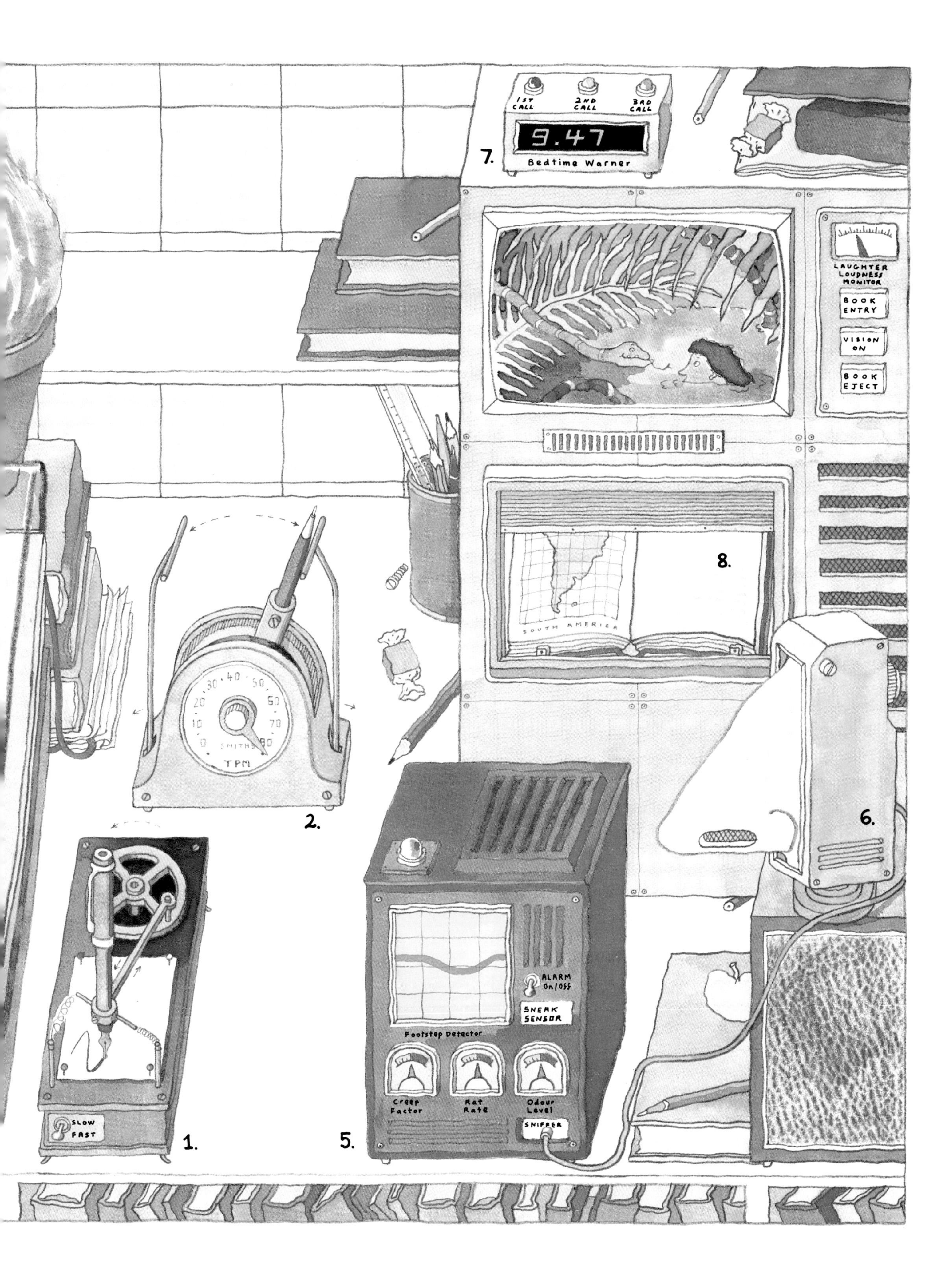

1ST CALL
2ND CALL
3RD CALL
9.47
Bedtime Warner
7.
LAUGHTER LOUDNESS MONITOR
BOOK ENTRY
VISION ON
BOOK EJECT
SOUTH AMERICA
8.
SMITHS
TPM
2.
6.
ALARM On/Off
SNEAK SENSOR
Footstep Detector
Creep Factor
Rat Rate
Odour Level
SNIFFER
5.
SLOW
FAST
1.

5.
4.
3.
6.
£1
2s
6d

GOING OUT

1. Disco Video Who's Who

The Disco Video Who's Who was invented in 1976 by Cynthia Cudgel, aged 14, for use at her local discotheque, Wiggles. With the special video display, Cynthia was able to scan the dance floor in secret, and zoom in on any person who caught her eye. A powerful micro-computer then gave her an instant data printout on the selected individual. Equipped with this information and a naturally ingenious mind, Cynthia was in a prime position to spread gossip, plan relationships and scheme schemes. Before long the entire social life of Wiggles was under her control. For five years she reigned supreme, until one day she accidentally pointed the Who's Who at herself. She was so insulted by the rude description it gave of her that she jumped on the machine, smashing it to bits. It was later given to the museum and re-constructed by technicians.

SPECIMEN PRINTOUT

NAME : TREVOR DONUT	
SEX : MALE	AGE : 14 yrs 5 mths
HEIGHT : 187 cm	WEIGHT : 101.5 kg
HAIR : MOUSEY	GREASE FACTOR : 87/100
B.O. LEVEL : 8.5	BREATH : NORMAL-SMELLY
PIMPLE COUNT : 597.2	DANDRUFF DENSITY : 9
FEET : CLUMPING	EARS : LARGE EYES : 2
BRAIN : + NO DATA AVAILABLE +	
GAIT : LURCHING	POSTURE : APE-LIKE
GIRLFRIEND : YES	PRESENT : YES
NAME : CHARLOTTE CABBAGE	
FOR FULL GIRLFRIEND ANALYSIS PRESS BUTTON B	
+ THANK YOU + HAVE A NICE TIME +	

2. Parent De-Irritation Chamber

Stops parents getting angry when you come in late.

3. Photon-Powered Rollerskates

This ridiculous and dangerous invention was the 'brainchild' of Sidney Splite, aged 5.

4. Remote-Controlled Clock

Makes parents believe it is earlier than it really is.

5. Parent Chauffeur Outfit

Natty outfit designed by Sandra Meek, aged 11, for her father. Mr Meek was also provided with a radio bleep, to summon him swiftly to his daughter's presence for late night collection.

6. Pocket-Money Extractor

An automatic parent up-ender, designed in 1959 by Clarissa Skinton-Broke, aged 12. The unsuspecting parent is induced into the Holdfast *PosiGrabbers* by the alluring smell of roast potatoes, and is rapidly up-ended into the cash delivery position. Deadly accurate rummaging arms quickly locate fruitful pockets, and a mighty 5000cc *Vibro* Jigger shakes the contents into a handy drawer.

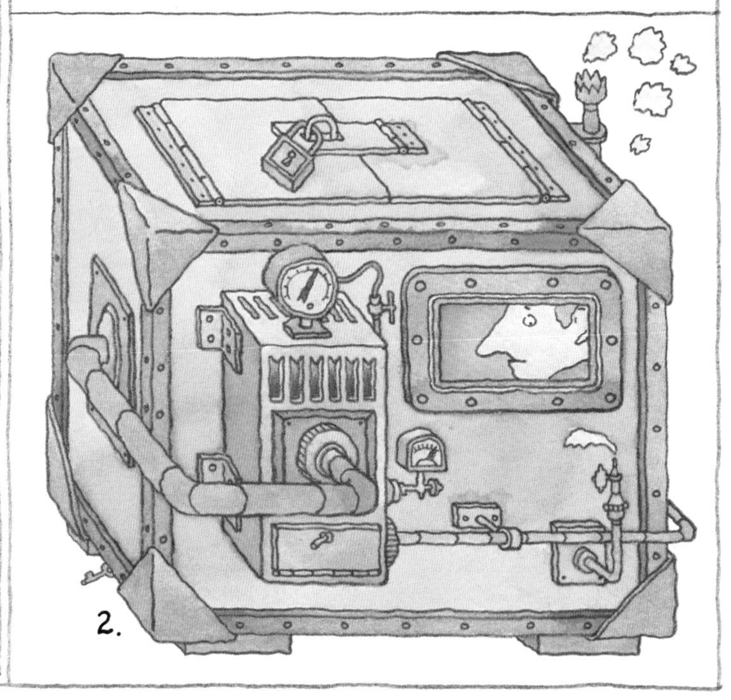

EXAMS
1. Lucky François Exam Mascot
Press beret button for instant grammar.
2. Spectacle Case Encyclopaedia
Invented by Myrtle Swatt, aged 14, who was large of brain but short of sight.
3. Clockwork Nailbiter
An early 19th century mahogany nail-biter with variable-speed gnawing action and handy clippings drawer.
4. Historical Pencils
These pencils were designed by Ivor Dodge and Cliff R. Cloggs, aged 13, for use in history exams. The boys went rapidly from bottom to top of the class, until an alert cleaning lady, not open to bribes, found the cause of success on the exam room floor.
5. Encyclo-Quartz Watch
Invented by Kevin Jones, aged 15, this watch converts at the touch of a button to a fully illustrated encyclopaedia.
6. Exam Gums
When all else fails. These gums contain ArGh₂, a deadly poisonous compound with no known antidote.
1.
LUCKY FRANÇOIS
2.
Blind & Bat (OPTICIANS) Ltd
These Spectacles belong to:
Myrtle J. Swatt
5.
CHRONO QUARTZ
SECS MINS
DAY DATE
MON
6.
The Original EXAMINATIO Throat Lozeng
PETE luvs SANDRA
SPURS

PANIC RATE
slow
fast
PRESS
3.
NOUN CALL
VERB CALL
TEACHER COMING
NOUS SOMMES
VOUS ETES
ILS SONT
ELLES SONT
-ETRE-
William IV [b.1765 r.1830-37]
Younger brother of George IV, third son of George III,
as the 'Sailor King,'
pular and eccentric.
Victoria [b.1819
Daughter of Ed
throne at the age
cousin Prince Alb
nine children. She
of Kent. She came to th
n 1841 she married h
as a devoted wife and
t Osborne House, I.o.W.
Question One
-1900
4.
LUCKY FRANÇOIS
1501-1600
PAGE 270352 — CORN LAWS — History, English.
and it was this proposal which formed the basis of the CORN LAW of 1815. In 1828 William HUSKISSON, Tory reformer, introduced a sliding scale of duties according to price, in order to relieve the distress caused by the high cost of
PAGE FORWARD
PAGE BACKWARD
INDEX FORWARD
INDEX BACKWARD
Help-U-Pass EXAM micro unit 7000H
ON
HOLD
Magnify x2
SNAP-SHUT
Display Light
SELF-DESTRUCT
ON
KEV
PETE
luvs
SANDRA
ENCYCLO QUARTZ
UP DOWN
START STOP
of cathode rays was first noticed by Sir William Crookes in the 1870s. It was a
-CATHODE RAYS-
INDEX
FAST SLOW FLIP
LEFT RIGHT
MAG x5
Throat Lozenge
EARLY POISONOUS
SPURS

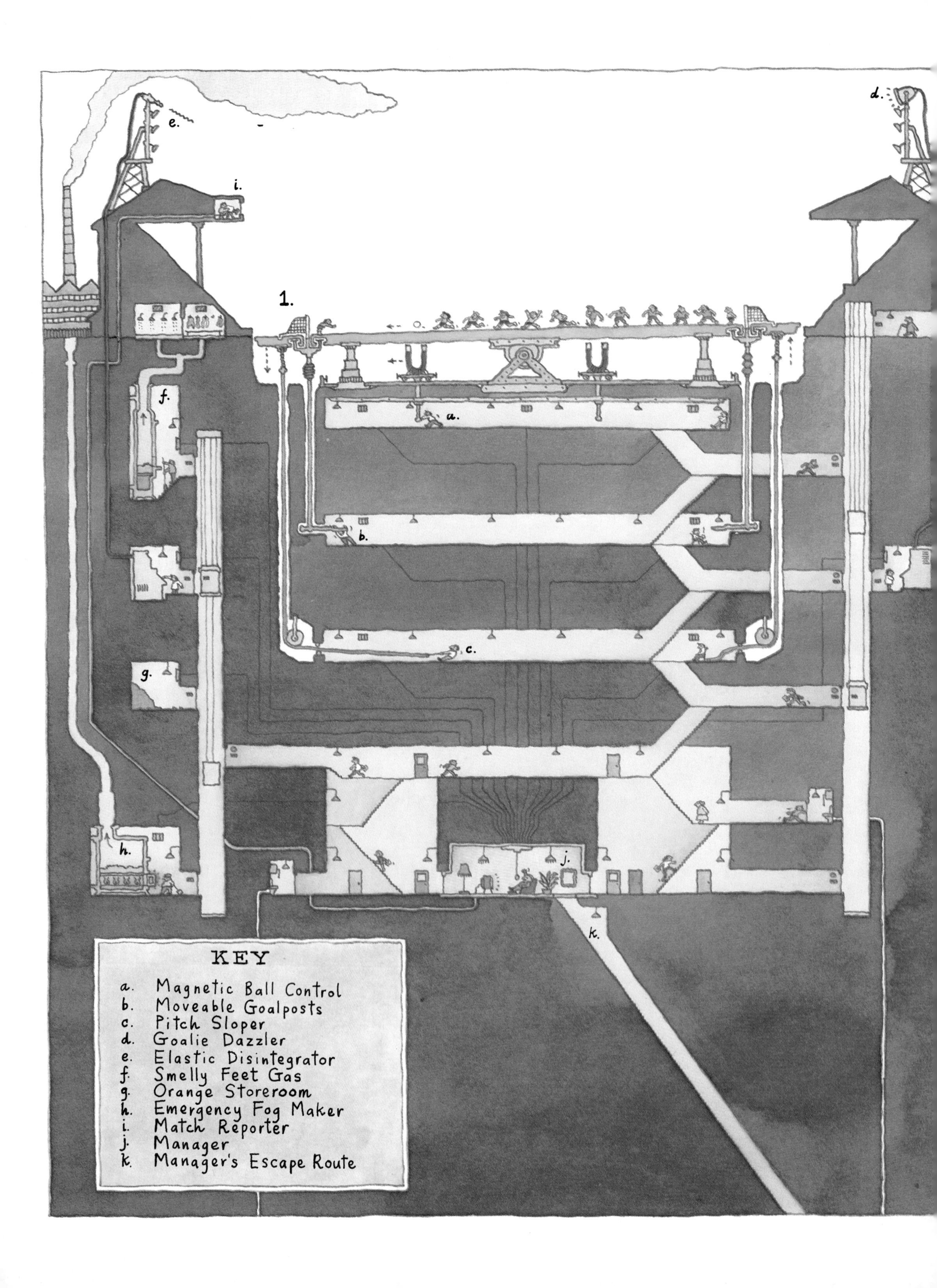
1.
a.
b.
c.
d.
e.
f.
g.
h.
i.
j.
k.
KEY
a. Magnetic Ball Control
b. Moveable Goalposts
c. Pitch Sloper
d. Goalie Dazzler
e. Elastic Disintegrator
f. Smelly Feet Gas
g. Orange Storeroom
h. Emergency Fog Maker
i. Match Reporter
j. Manager
k. Manager's Escape Route

SPORTING INVENTIONS

1. Remote-Controlled Football Stadium

This is a drawing from the museum's fine collection of rare sporting prints. It shows a scheme for controlling play during football, and was proposed in 1957 by D.R.T. Fowles, aged 12, for his home team Hardleigh Athletic. The team manager sat in an underground chamber watching live action coverage of the game above on TV. Through a microphone and a system of loudspeakers, he could shout instructions to technicians in various departments. Each department housed a device designed to influence the course of play. Unfortunately for Hardleigh Athletic the scheme was never adopted, and after a ten year losing streak the club was forced to close. The stadium is now the site of a new leisure complex.

NOTES

1. The Elastic Disintegrator produced a powerful beam which rotted elastic in football shorts. Hardleigh Athletic were to be protected by having shorts held up by string.
2. Smelly Feet Gas was a thick, nauseous vapour pumped through air vents into the opposing team's changing rooms. It was claimed to make the victims foolish and clumsy of foot.
3. In the Orange Storeroom were kept the spiked half-time oranges.
4. The Emergency Fog Maker was designed to blot out all vision if things went wrong and the opposing team looked as if it were winning.
5. If things went badly wrong and the opposing team won, the manager could protect himself from his own team by locking the doors of his armour-plated room and escaping down the back stairs.

2. Go-Faster Swimming-Trunks

These useful swimming-trunks were invented by Donald Snorkle, aged 9, and were powered by two tiny engines disguised to look like Olympic Medals. Each engine converted chlorine in swimming-pool water to polyzilgooxymoronate [$Zo_2O_3Om_4$], a powerful jet propulsion fuel.

3. Sports Kit

Contains a gym greasing gun, selected verrucas, jockstrap shrinking spray, and specially bred, voracious tennis-racquet woodworms.

4. Magic Football Scarf

Changes colour instantly to suit the occasion.

5. Cross-Country Run Comforter

An all-over thermal suit for those bleak, wintry cross-country runs. Made from durable Plodalon, with armpit ventilation.

THE *Automatic* PESTERWHEEL OF MONTMORENCY SNOUT

"For sixteen weeks Snout hammered and sweated under the lights in his yard, pausing only for refreshment from a flask of nettle tea. Now, with a spanner in his hand and the yellow light of malevolence in his eye, he pushed the heavy brass lever to 'forward roll'. Slowly, inevitably, with a noise like the rumbling of a thousand municipal street-cleaners, the Pesterwheel began its first disturbing rotations . . ." *

The Automatic Pesterwheel of Montmorency Snout was probably the nastiest of all the Public Nuisance Machines on record in the museum's archives. It was more bothersome by far than the Hovering Vexerator of Enid Skrunk; it plagued more people than the Guided Boldergrab of Greville Boilz; it was even more irksome than the Compost-Powered Flostersnoddy of Daphne Meadowsweet.

The Pesterwheel rolled along streets, sucking up fires from people's chimneys to heat its massive steam boilers. Each side of the Wheel were chimney-pot bonking rods and dangling window biffers. There were huge rubber hoses that sucked up stuff from the drains, and sprays that squirted it out again on pavements and people's doorsteps. There were flowerbed tramplers, tile dislodgers, milkbottle scatterers, and hydraulic lamp-post tweakers. Lights dazzled, speakers blared, and vast fans blew out litter by the ton. There was a garden gnome snaffler, a pet snitcher, and down in the bilges were rusty compartments crammed with stray cats. Snout's Wheel could mangle railings, chew bicycles and crunch up paving-stones like nobody's business. Its unpopularity was enormous, and it was given an International Nuisance Factor Rating of 99.98 by the visiting inspectors.

The picture shown here is an artist's impression based on old photographs in the museum's collection.

*from 'The Life and Triumphs of Snout the Great' by Montmorency Snout.

SEMPER MALEVOLENS

GLOSSARY

Chris Winn and I have attempted to recreate what we observed as faithfully as possible. But we have not illustrated everything we saw. Some things we didn't understand, other items even we considered downright disgusting and some which didn't need illustrating since they were self-explanatory. But to keep our promise of making this as complete a catalogue as possible here are three lists of the missing inventions.

1. INCOMPREHENSIBLE

Electric Metric Snime Timer — 'snime' ?
Thin Bore Piddle Pipe — possibly rude
Razor Sharp Fluming Irons — seemed uncomfortable
Pre-Timed Snargle Tweezers — appeared painful
Advanced Sog Distribution — sounded messy
Adjustable Godgering Rings — looked very sinister

2. OBJECTIONABLE

Zit Squidger — positively a health hazard
Puke Dispenser — messy proof of bogus illness
Nose Gunger — creates useful sound of heavy cold
Fake Can Label — place around dog and cat food tins and serve to boring relatives

3. SELF-EXPLANATORY

Automatic Crooked Card Shuffler — guarantees you always get the best deal

Homing Marbles — never lose one again (suppressed by the Marble Manufacturers Society)

Nightmare Eraser — a boon to vivid imagination

Silent Hair Tangler — horrific moments when victim tries to comb her hair

Stringed Coins — retrievable, for use in vending machines
Turbo-Driven Shoe Cleaner — outlawed by shoe-brush manufacturers
Sneezing/Itching Powder Blowpipe — guaranteed confusion (victim doesn't know whether to blow or scratch)

Gum Gun — for firing stale wads of chewing-gum with deadly accuracy
Paper-Thin Non-Melting Chocolate — place inside school textbooks (appears to be just another page)

Everlasting Supper Drink — abandon all fear of the dreaded parental 'As soon as you've finished your cocoa off you go to bed'

PRINTED IN BELGIUM BY
proost
INTERNATIONAL BOOK PRODUCTION

3267E